AWESOME
ANIMAL EVOLUTION

Dr Nick Crumpton

illustrated by

Viola Wang

Introduction

No matter where you look on the tangled branches of the Tree of Life, from brainless sponges to super-intelligent primates, or from tiny insects to enormous elephants, every animal has to solve lots of problems just to make it through the day. If you live under the ground, you've got to move a lot of dirt and soil out of your way. And if you spend your days swinging through the branches of tropical rainforests, you'd better be great at staying up there!

Brilliantly, animals' bodies have changed slowly through thousands of generations and over hundreds of thousands of years so that their **BODY PARTS** help them do the jobs they need to do. Each generation, the animals born with body parts that help them solve these problems more easily — *slightly* longer necks that make it easier to reach food, legs that run *slightly* faster to out-sprint predators — have babies and pass those body parts and skills on to the next generation. That's **EVOLUTION**!

The reason *why* animals all look so different from each other is because of what they do — how they catch food, where they live, which predators they need to escape.

Have a look at the Tree of Life on this page. The further away an animal's branch is from another, the more *distantly* they are related, and, usually, the more different they look! Reptiles and bony fish are quite closely related (they both have skeletons, similar-looking brains and eyes that work the same way), but sharks and jellyfish are hardly related at all — and they look very different from each other.

But evolution has been *amazing* at equipping animals with the *same* parts and skills to do the same jobs, no matter where they live or what they are. Some of those parts and skills are *so* helpful and *so* useful, that they are found in animals all over the world. They've evolved in animals that are totally unrelated to one another and even in animals that lived *millions* of years apart. These are the animal **SUPER POWERS** and wonder parts!

Take another peek at the Tree of Life. The stars on the tree

JELLYFISH✶

ACORN WORMS

SPONGES

ORIGIN OF ALL ANIMALS

show where one super power evolved once, and then again, and then even more times completely *independently* on unrelated branches (you can find out what it is on page 16). But there are many more examples! So let's dive in and get to know the triumphs of evolution: the super powers that turn up in the Tree of Life time and time again (and again ... and again!).

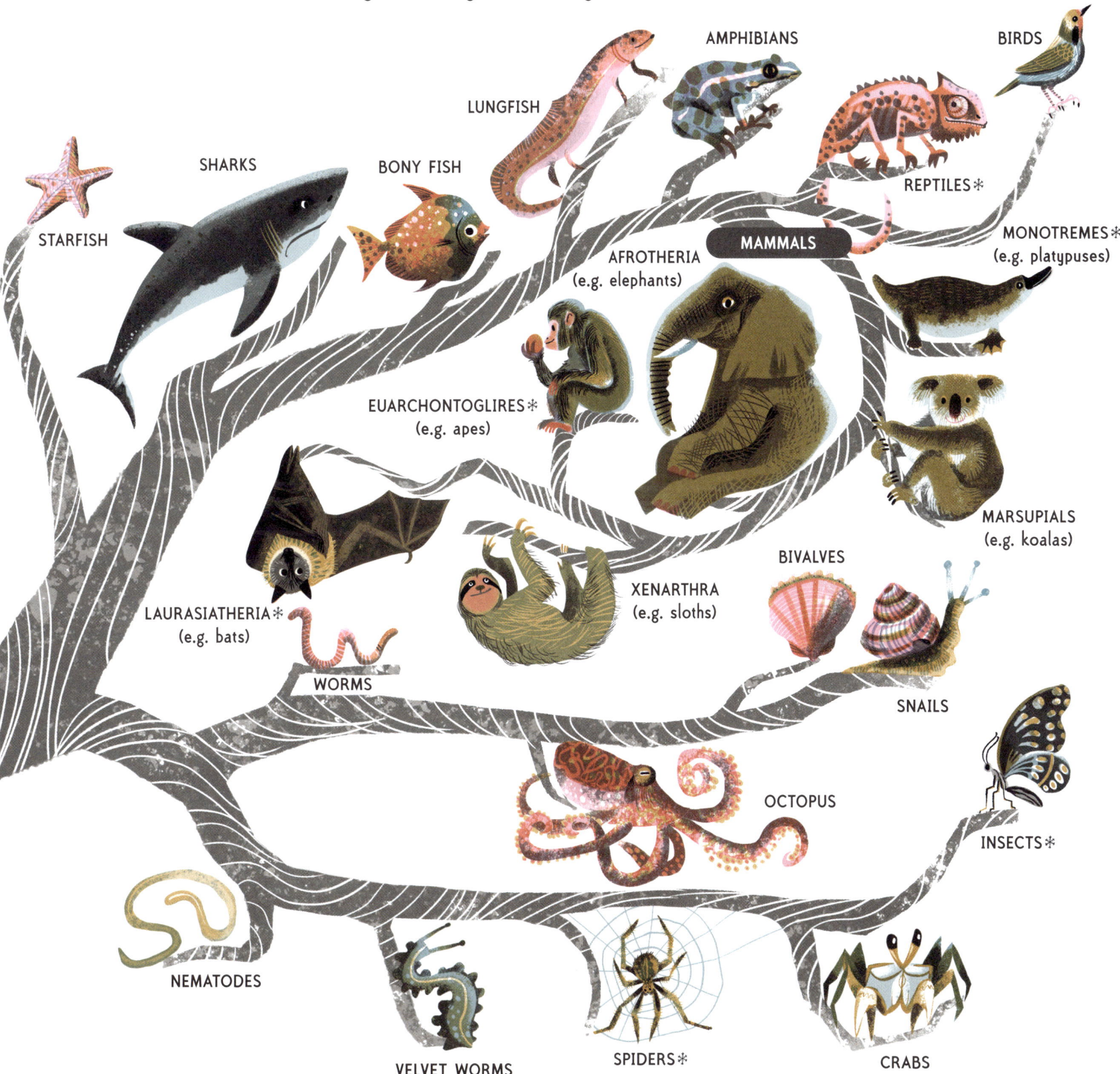

Body Armour

If you picked any animal on Earth at random, no matter where it lives, no matter what it looks like or what it smells like, it's pretty likely that something else wants to eat it.

So most animals have to live their lives avoiding becoming another species' lunch. Hey, it's dangerous out there! But an incredibly large number of species have all evolved <u>the exact same way</u> of protecting themselves — body armour.

For three billion years, all life was tiny and squishy, and at some point a few of these squishy shapes began eating others. But then, 750 million years ago (that's 500 million years before the very first dinosaurs appeared), some small creatures began to protect themselves by evolving armour. This was an unbelievably important step in the history of how-to-escape-being-eaten!

This super skill, which scientists call **BIOMINERALIZATION**, was so helpful that over the next three quarters of a billion years, lots and lots of animals evolved protective armour...

→ *Cloudina* lived long ago, before the main types of animals we know today first appeared. Its cone-like casings protected the soft animal inside from some of the earliest predators. Snail-like shells also evolved at around the same time as *Cloudina*. Today, there are over 80,000 species of shelly **molluscs**!

↑ **Turtles** evolved their armour 260 million years ago. Their shells are made out of lots of bones that are usually found on the insides of reptiles, like the ribcages. This means that turtles' shoulders and hips are actually protected *inside* their ribs!

The animals that build **corals** are small and soft. They produce skeletons around them, which act as incredibly tough protection against the battering of the sea and also the sharp beaks of parrot fish. →

Ankylosaurus

Komodo dragons have armour that is made from thousands of tiny pieces of bone which grow within their skin, underneath their scales, called osteoderms. These act as an almost impenetrable suit of chain mail protecting them from other Komodo dragons when they are battling over food and mates.

Other, extinct reptiles, such as armoured dinosaurs like **Ankylosaurus**, also protected themselves by evolving osteoderms ... even though dinosaurs are more closely related to sparrows and chickens than to lizards like Komodo dragons!

Body armour is really rare in mammals (animals that have backbones and fur and produce milk), but **pangolins** rely on their coats of scales to protect them from predators like lions. Their scales are made from keratin — the same hard material that makes up human fingernails and rhinoceroses' horns. ↓

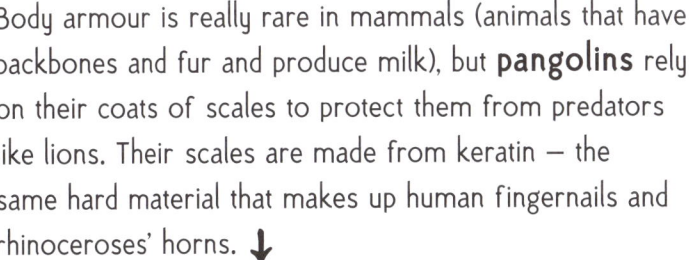

Komodo dragon

Osteoderms

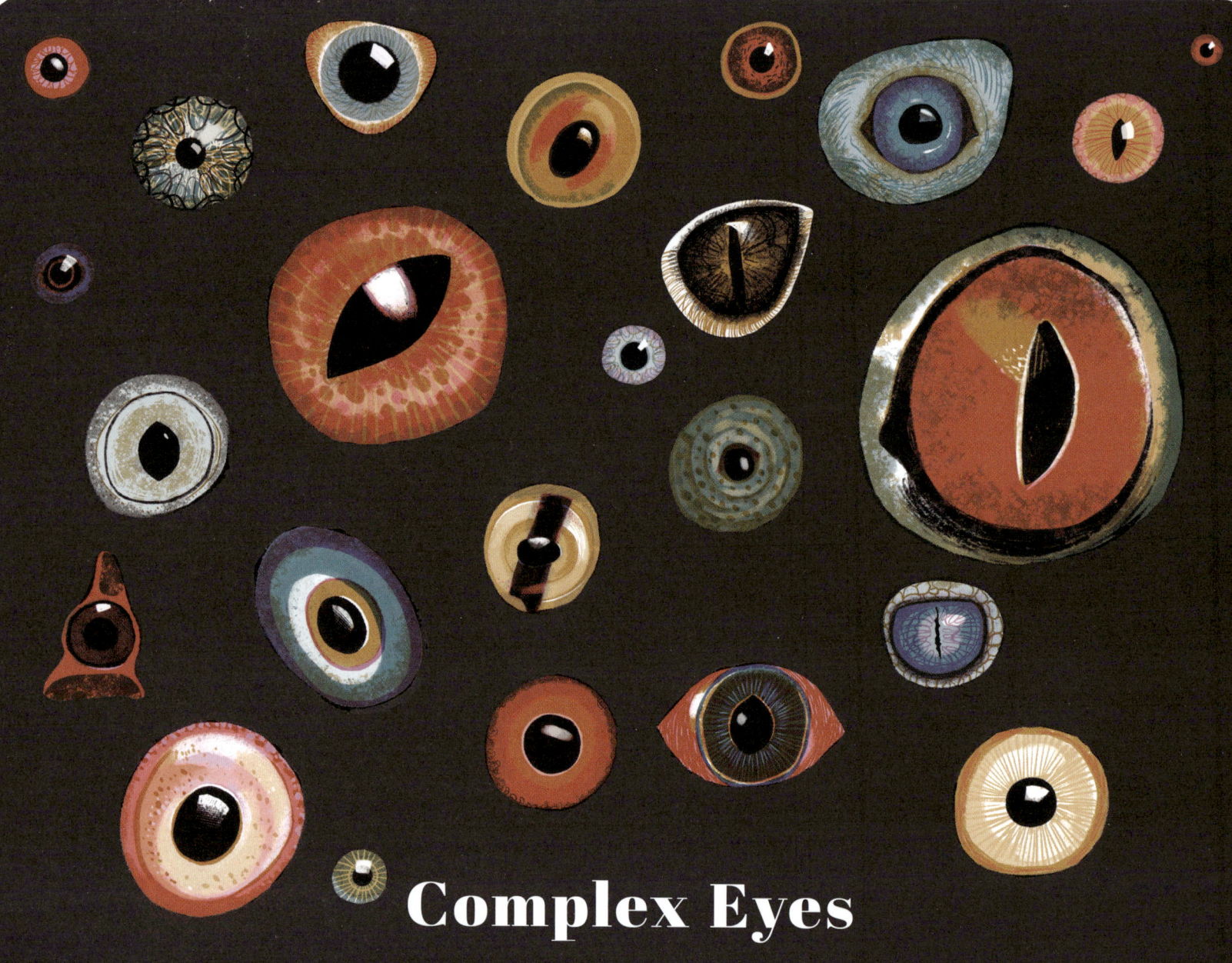

Complex Eyes

There are lots of ways to sense what the world around you is like. Some animals depend on touch, while others use their sense of smell to work out where to find a tasty spot of lunch, or to avoid becoming something else's dinner.

Most senses work best when you are quite close to whatever you are trying to detect. But if you can capture light after it bounces off something and turn it into a signal to your brain, you can **SEE** from a distance. The simplest eyes (which could only detect light and dark) first evolved an incredible 600 million years ago, when life was only found in the oceans. Today, a huge variety of life can see, even if only with these simple eyes.

Sensing light is all well and good, but the better you can see, the better you can spot danger or a snack, <u>and</u> the more easily you can communicate with other animals using signs and colours. And to see better, you need a more sophisticated set of complex eyes — which is exactly what evolution gave to a whole bunch of animals!

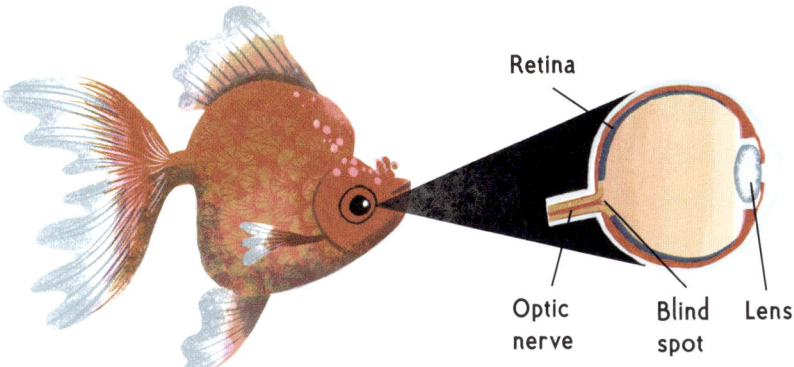

↑ Vertebrates, which are animals with backbones (that includes us!), evolved more than 500 million years ago. Today, all vertebrates that have eyes, from **goldfish** to gorillas, have the same complex "camera-type" eyes, which precisely focus light through a clear lens onto a special, sensitive area at the back of the eye called the retina.

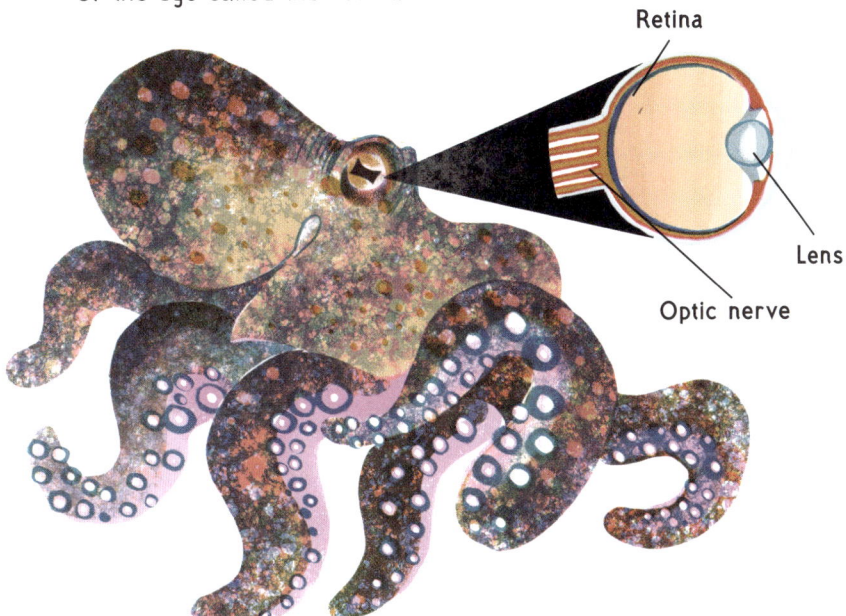

↑ On the other side of the animal kingdom from vertebrates, **octopuses** evolved similar camera-type eyes completely on their own. Their camera-eyes have lenses and a sensitive retina, just like vertebrates, but the outside of their eyes aren't covered by a protective layer, like ours. Octopus eyes work even better than vertebrate eyes. Unlike us, they don't have blind spots because of the more efficient way that their eye nerves are attached to their brains!

↓ **Jellyfish** are about as distantly related to vertebrates as it's possible to be, but at least one — the box jelly *Tripedalia cystophora* — has also evolved a camera-type eye. Two camera eyes hang from each of the four corners of the jelly's bell. Although a jellyfish doesn't have a brain as such, the light signals are sent to a net of nerves that covers the bell, which then "decides" what to do with the information!

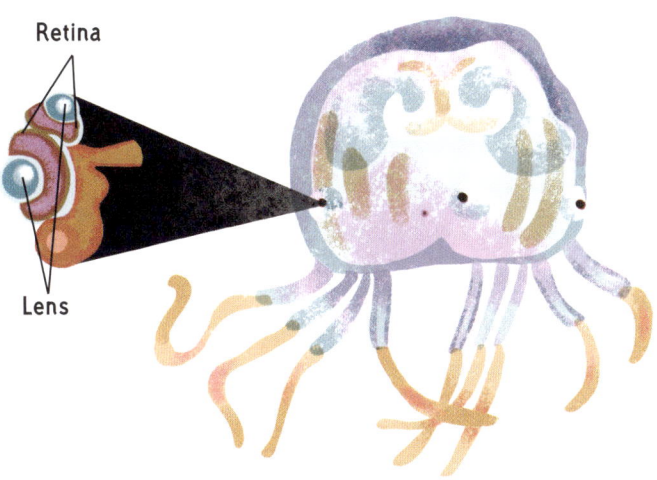

↓ Sight evolved in a completely different way on yet another branch of the Tree of Life. **Insects** have compound eyes. Rather than focusing light through one lens, insects rely on thousands of very small hexagonal parts called ommatidia that sense light and colour. The ommatidia are grouped together, which allows insects to see in almost every direction at once.

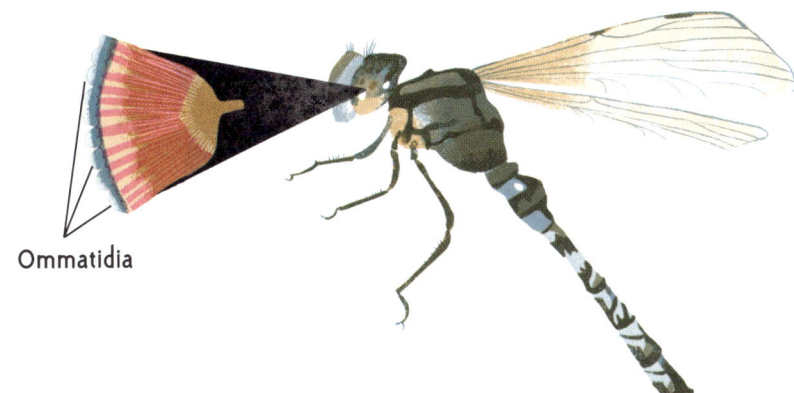

← The extinct **trilobites**, who lived on the sea floor for 250 million years before even the dinosaurs appeared, also had compound eyes. But these tiny eyes were built from calcite crystals — they had eyes made out of stone! These crystals were transparent, so acted just like clear lenses.

Wings

The ground is a perfectly fine place to live. It has places to hide, animals to hunt and plants to nibble. But those hiding places can be discovered, those animals might be too quick to catch, and there are always taller plants just out of reach. So it's not that surprising that flapping **WINGS** evolved really, really quickly after life first stumbled out of the seas.

Being able to fly is a great way to escape predators and to travel big distances much more quickly than walking. It's a really efficient way of getting around.

But it's also very complicated to fly with flapping wings. It requires large, curved wings to create "lift", an almost magical way of being suspended in the air. It also needs massive amounts of muscle power to keep pushing the animal forwards.

Despite this, flying using flapping wings has been "discovered" many times during the last 400 million years...

→ Early **insects** and their relatives were the first animals to take to the sky by beating their wings. Zoologists (scientists that study animals) still aren't *exactly* sure how their thin wings evolved — they might once have been gills, or maybe they were originally used to help insects warm up or cool down. The griffinfly ***Meganeuropsis*** lived 290 million years ago and was the biggest flying insect ever, with a wingspan of 70 centimetres. That's as big as a sparrowhawk!

Pterodactyl

Meganeuropsis

↑ Most insects actually use their wings in a slightly different way from the other animals shown here. They sweep their wings *backwards* and *forwards* as well as up and down. Some mosquitoes today beat their wings nearly 800 times per second!

→ At some point in the Jurassic period between 201 and 145 million years ago, a few lucky feathered dinosaurs such as the **Archaeopteryx** were born with slightly chunkier chest muscles and lighter bones than their relatives. Feathers up to this point had just been used to keep dinosaurs warm, but these stronger muscles let them take to the sky, using their light feathers to help them catch the air. Birds are the only dinosaurs still alive today. Take a look at a roast chicken's skeleton the next time someone cooks one and you'll be able to spot your dinner's inner *Velociraptor*.

Archaeopteryx

Rather than using feathers, **bats** use thin skin stretched between their long fingers to push down against the air as they fly. Being the only mammals that can truly fly has meant they can eat insects that no others can reach, which has helped them become super successful. Today, one fifth of mammal species are bats! →

Bat

↑ The only other animals to have evolved flapping flight were the **pterosaurs**. These reptiles, which were only distantly related to dinosaurs, lived around 220 to 66 million years ago and discovered flight completely on their own. Like bats, they stretched the skin of their wings out from their hands, but they held their wings on a very, very long little finger — a beautiful example of evolution coming up with exactly the same solution to a problem, but in a completely different way...

Eating Ants and Termites

You might have noticed by now that finding things to eat is <u>really</u> important to animals. And once they've found food, it's also important for them to make sure it isn't snaffled by other animals.

Some animals become experts in eating one particular kind of food that no other animal can eat. The food might be hard to reach, toxic, or just taste bad. If you're the only animal that wants to eat a food, then you've got it all to yourself!

It would be even better if the animals that you wanted to eat were everywhere — and there aren't many animals that are found in greater numbers than insects: specifically, **ANTS** and **TERMITES**.

The number of ants in the world is mind-blowing. There are probably 20,000 different kinds (in comparison, there are only about 6,400 types of mammals, from shrews to whales). They are found almost everywhere apart from Antarctica, and all together probably number over 100 *trillion* individuals. Termites don't live in as many places, but as social insects, they are found in enormous numbers inside their mounds.

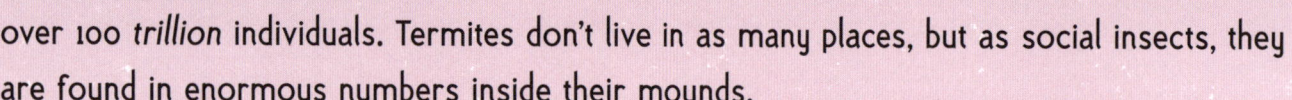

Eating ants and termites isn't easy: they're small, live underground, *hate* being disturbed and pack powerful bites and stings. Which is why the animals that have evolved to eat this secret bonanza all share a few extremely useful tricks.

↓ The **giant anteater's** tongue is an enormous 60 centimetres long. It's a third of the length of its body! That's one of the longest tongues relative to an animal's size in the whole animal kingdom and is perfect for reaching into its favourite snack's nests.

Giant anteater

Having powerful forearms which are too small for burrowing is usually a giveaway that an animal has evolved them to break into insect mounds. That's why palaeontologists (scientists who study extinct creatures) think that the dinosaur *Shuvuuia* might have been an ant-eater about 75 million years ago.

Aardwolves might look like they hunt large animals in packs, like their close relatives hyenas, but in fact they use long, sticky tongues to catch and eat insects, including ants and termites.

Anteaters and **tamanduas** from Central and South America are closely related to sloths — some of them even live in the branches of trees. They slurp up ants with no trouble thanks to their incredibly long tongues covered in sticky saliva. They also have large claws to tear open anthills and termite mounds.

Those powerful arms and long sticky tongues are very similar to those found in **pangolins** from Asia and Africa. Scientists used to think pangolins were very closely related to anteaters, but we now know that they are more closely related to lions, pigs and even whales than to their sticky tongued lookalikes. They have also evolved scaly armour to protect them from the ants' defensive bites.

Shuvuuia

Aardwolf

Pangolin

Tamandua

Seeing with Sound

For a very long time, humans were totally confused about how bats managed to see so phenomenally well at night with only minuscule eyes. So well, that not only do they avoid crashing into things in the pitch dark, but they can also capture insects just millimetres long.

It was only in the 1930s that scientists finally proved that bats were using both their incredible sense of hearing and an amazingly clever super skill named **ECHOLOCATION**. By shooting out pulses of high-frequency sounds (much higher than humans can hear) and listening to how long faint echoes of the noise take to bounce back off objects around them, animals that echolocate can create a super-accurate map of what's around them.

At the time of this discovery, humans were still developing "sonar" — a technology used by submarines to silently detect other boats that works in exactly the same way as echolocation. Lots of people couldn't believe evolution had beaten the world's smartest engineers to it!

But it isn't just bats: a whole load of other animals also evolved this super power to help them "see" when their eyes would be useless.

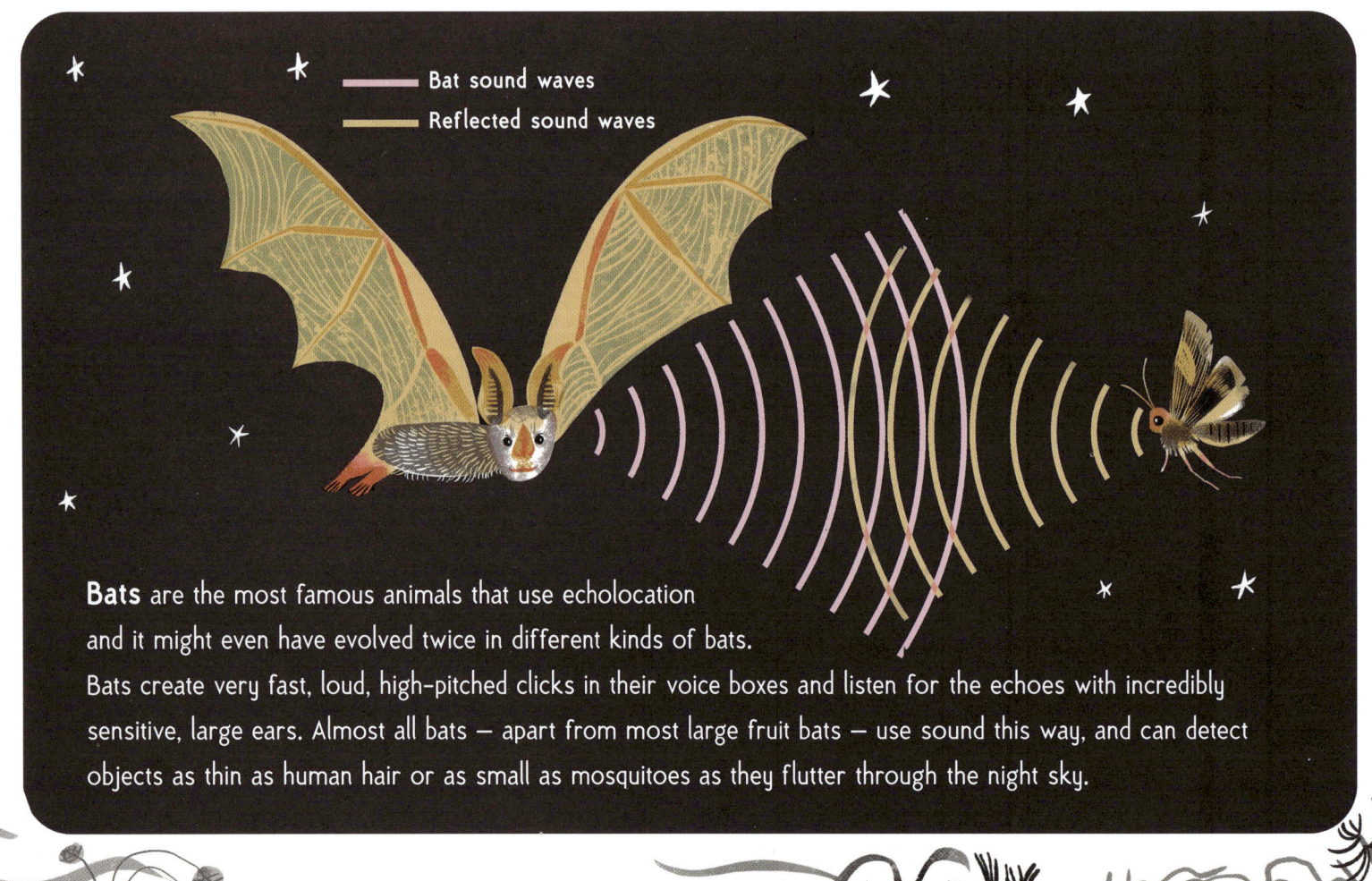

— Bat sound waves
— Reflected sound waves

Bats are the most famous animals that use echolocation and it might even have evolved twice in different kinds of bats. Bats create very fast, loud, high-pitched clicks in their voice boxes and listen for the echoes with incredibly sensitive, large ears. Almost all bats — apart from most large fruit bats — use sound this way, and can detect objects as thin as human hair or as small as mosquitoes as they flutter through the night sky.

The other common echolocating animals are toothed whales like **dolphins** and orcas. Although they have come up with the same way of finding prey as bats, they make sound very differently. A toothed whale creates clicks from its "monkey lips" — a set of muscles just behind its blowhole — and focuses them through the bump on the front of its face. Rather than listening out for echoes with large ears, it receives sounds through its lower jaw, which is full of fat and directs the sound back towards its ears.

Monkey lips

Some birds that hunt prey in the dark or roost in pitch-black caves have also evolved echolocation. Both Asian **swiftlets** and South American oilbirds make clicks in their vocal cords in a similar way to bats, in order to avoid flying into things in the dark.

It's not just animals living in dark caves and murky water that use echolocation. When scuttling around on forest floors and along branches, **soft-furred tree mice** found throughout China use high-pitched "twitters" to find dimly lit hiding places and escape routes from predators. Even though they live in very different environments and aren't very closely related, these mice, just like dolphins, swiftlets and bats, have evolved brains that can make sense of the echoes they receive and turn them into a mind-map of their surroundings.

Venom

Being eaten is the worst. No animal on Earth (apart from a few parasites) wants to be another's dinner, and so animals have, over millions of years, evolved tens of thousands of ways to avoid being nibbled.

Lots of animals simply run (or swim, or hop, or jump, or squelch) away from predators — but that uses up a lot of energy. And what if those predators can move faster?

A much more elegant solution is to make yourself taste horrible, so no other animals want to eat you. Loads of species create poisons using chemicals in their bodies to make them taste awful; some of these can even make their attackers sick. Being <u>toxic</u> is a brilliant way to fend off would-be attackers and is a strategy found all over the animal kingdom.

Another, more devious use for these poisons is to attack other animals with them, and stop them in their tracks so *they* can be eaten! And the best way to get poison into another animal is by injecting it with a sharp tool like … say … a sting! When animals inject toxins into other animals like this they are called **VENOMOUS** and that duo — poison and a sharp, pointed part — has evolved time and time again.

↓ **Snakes** evolved venom in their saliva sometime during the Jurassic period and many types of snakes have since evolved lots of different sorts of poison. The venom is injected when the snake bites its prey, the toxin flowing either through hollow fangs, or down a ridge on the outside of the teeth.

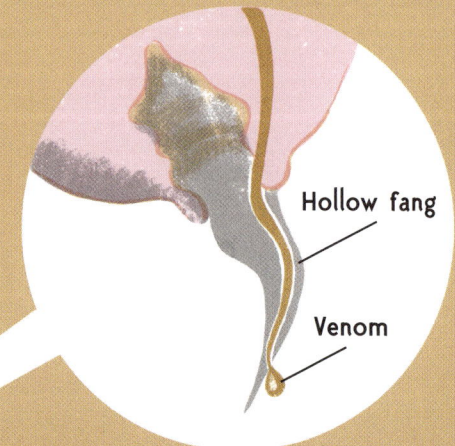

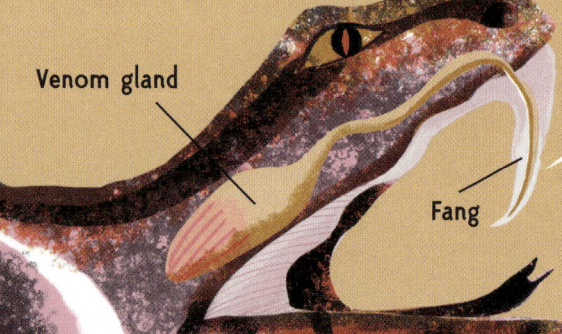

And what venom some have evolved! One of the most toxic snakes in the world is the inland taipan from central east Australia, whose bite contains enough venom to kill 100 humans!

Venom gland

← Venom also evolved in a few mammals, although this is very unusual. The rather innocent-looking **slow lorises** from South East Asia produce a venomous saliva that they use to attack other lorises when fighting for mates. Another venom also seeps out of a gland in their forearm, and they coat their fur with it to keep predators at bay.

→ **Solenodons** are rare, small mammals that live in the Caribbean. Beneath their long noses, which they use to sniff out prey under forest leaves, are two long incisor teeth with a groove for injecting venomous saliva — just like snakes. The only difference is solenodon fangs are in their *bottom* jaws, not the top.

← You don't need to have a backbone to pack a punch. Many **centipedes** are venomous, injecting their venom into their prey through forcipules, which are fang-like legs near their heads. Meanwhile, the Australian box jellyfish (yep, that's the same type of jellyfish that evolved camera-type eyes) is the most venomous animal in the oceans. Its long tendrils are covered with cells called nematocysts that contain coiled-up stings. These shoot out at high speed, injecting painfully potent venom into its prey.

Eating Dead Things

Planet Earth is home to an almost incomprehensible amount of life.

In just a handful of soil there may be thousands of insects and other invertebrates (animals without backbones), not to mention tens of thousands more microscopic creatures. And that's just one clump of dirt. There might be anywhere between ten million and one hundred million different sorts of animals living on the planet, and each year hundreds more species are discovered by scientists. But one thing unites them all — they all die.

That might seem depressing, but it means that for animals that just love to eat dead, rotting things — which is known as **SCAVENGING** — Planet Earth is completely covered in food!

Although it might not sound too appealing, scavenging is a behaviour found all over the animal kingdom — which is good news for the rest of us who don't want to be walking around knee-deep in dead things. All this nibbling, chewing, crunching and munching means dead things eventually transform into the soil and materials that make up the very ground that we walk on. Which is why it's great this super skill has evolved many times.

Sexton — or "burying" — **beetles** can sniff out dead animals from kilometres away and, after burying them in the soil, use them as nurseries to feed and protect their young. Unusually for beetles, they stay and look after their young when they hatch. They might live inside dead animals, but they're devoted parents. ↓

Sexton beetle

Beetle larvae

Old World vultures, found in Africa, Europe and Asia, are super important in their environments. Not only do they get rid of rotting carcasses by eating them, but they also destroy the bacteria in them that would be harmful to other animals if they ate any of the putrid meat.

One of the most amazing examples of animals from opposite ends of the world coming up with the same solution to an evolutionary problem is the **New World vultures**. These are a group of birds that, with their reaching necks, tearing beaks and bald heads, look almost exactly the same as the Old World vultures, but aren't related to them at all! Old World vultures are closely related to eagles and hawks, whereas the New World vultures evolved from the same branch of the bird family tree as the long-legged storks.

New World vulture

Old World vulture

Blowflies

Striped hyenas also act as part of the clean-up crew in the same environments as some vulture species. Just like vultures, they have very powerful stomach acid, which helps destroy bacteria that would be dangerous to most other animals. ↓

When it comes to getting rid of dead animals, insects have taken things to another level. Many species of **blowflies** lay their eggs in dead animals so that their larvae (which you might know as maggots) have an almost endless meal upon hatching!

Eggs

Maggot

Long Necks

Long necks have turned up so many times in the history of life on Earth — and in so many different sorts of animals — that they must be a really useful body part to evolve.

But, sometimes, working out what the parts of an animal are *for* is tricky. This is because body parts can have lots of different uses, not just one. And just because an animal uses that bit of their body to do a certain thing, it doesn't mean that's *why* it evolved in the first place. Think about your nose: our ancestors didn't evolve noses to balance their sunglasses on, but an alien studying humans might think that is why we have pointy bits of cartilage on our faces.

So although we see certain amazing animal parts showing up again and again, sometimes we're not sure why. And that's exactly the problem with **LONG NECKS...**

It's usually thought that **giraffes**, the tallest animals on Earth today, evolved their long necks in order to be able to munch on taller trees as climate change made the forests of their home disappear and they had to compete to eat the fewer, taller trees in the savannah. But male giraffes also use their long necks to battle each other and show who's the most powerful. Giraffes got their long necks by growing longer neck bones, rather than making more of them. They have the same number of bones in their necks as you have in yours! →

The necks of swans, **flamingos** and other birds are long because they have many more bones (called vertebrae) in their necks. They use these long necks to get to hard-to-reach food, like small animals and plants on the bottom of streams and waterways, without getting their entire bodies wet. →

Flamingo

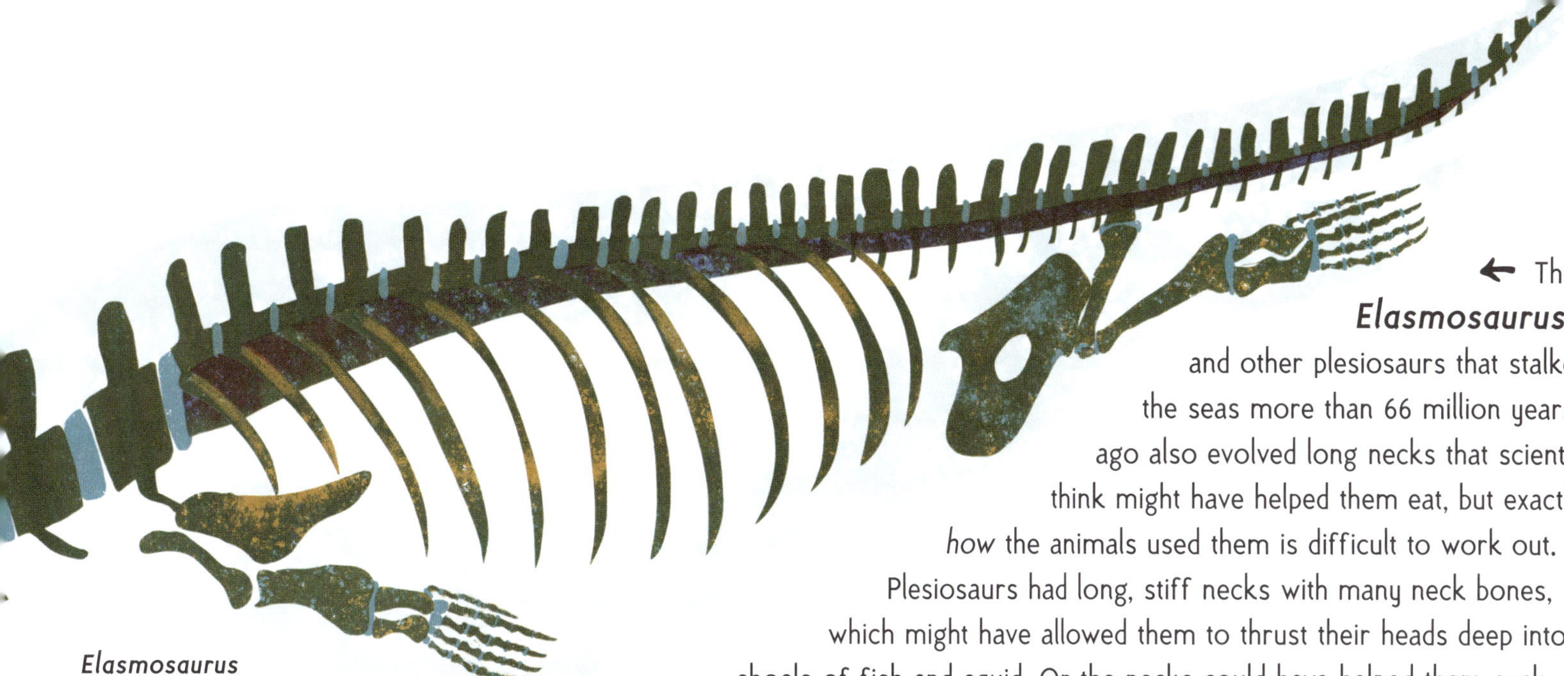

← The ***Elasmosaurus*** and other plesiosaurs that stalked the seas more than 66 million years ago also evolved long necks that scientists think might have helped them eat, but exactly *how* the animals used them is difficult to work out. Plesiosaurs had long, stiff necks with many neck bones, which might have allowed them to thrust their heads deep into shoals of fish and squid. Or the necks could have helped them suck food up from the seabed without scraping their bodies along the stony surface.

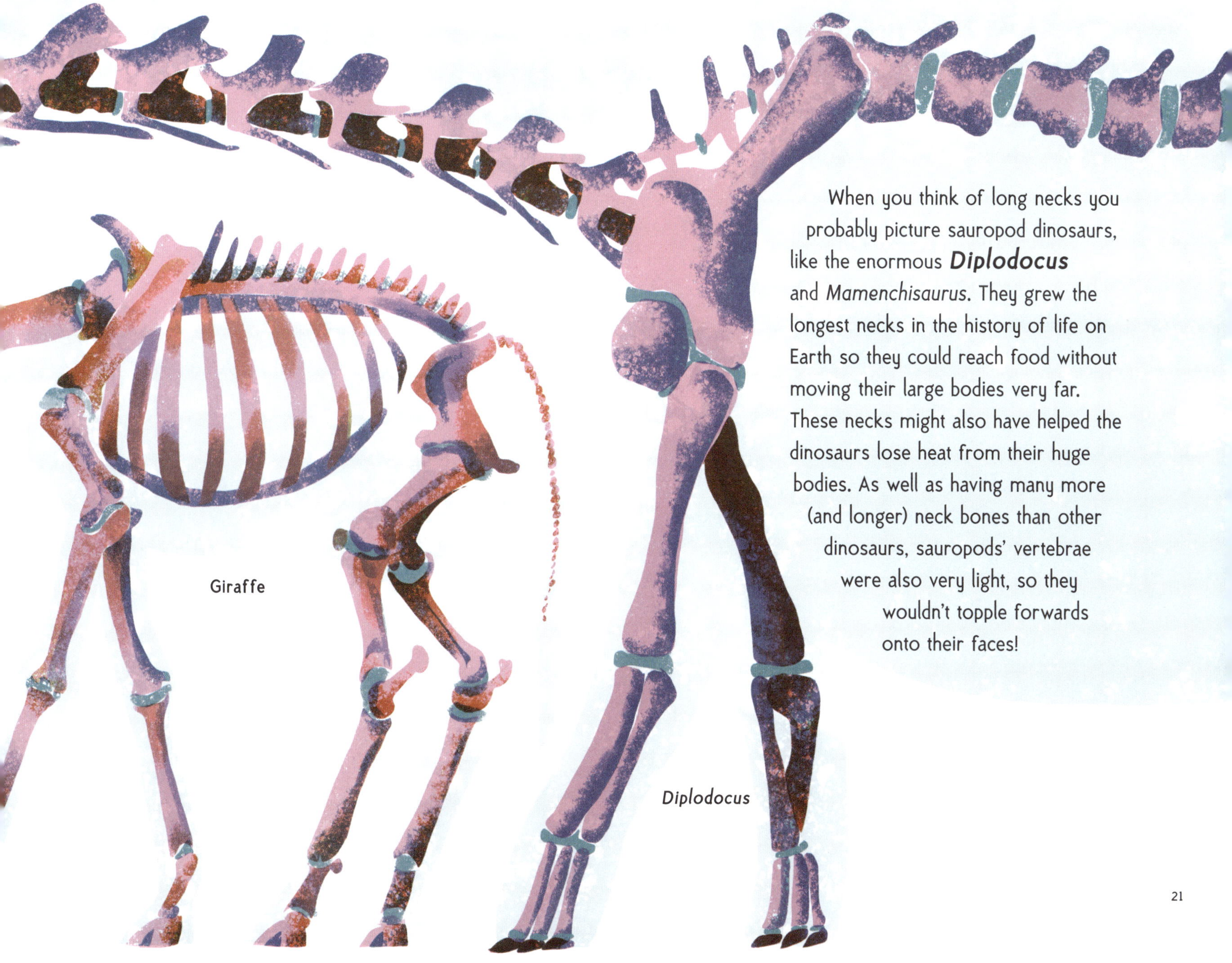

Elasmosaurus

When you think of long necks you probably picture sauropod dinosaurs, like the enormous ***Diplodocus*** and *Mamenchisaurus*. They grew the longest necks in the history of life on Earth so they could reach food without moving their large bodies very far. These necks might also have helped the dinosaurs lose heat from their huge bodies. As well as having many more (and longer) neck bones than other dinosaurs, sauropods' vertebrae were also very light, so they wouldn't topple forwards onto their faces!

Giraffe

Diplodocus

21

Jumping

When it comes to escaping danger, scampering away generally means sprinting forwards or backwards, left or right. But what if your predator can scamper after you too?

Many animals have come up with a similar solution to avoid trouble — they've found the safest way is <u>up</u>!

JUMPING is an *incredibly* old skill and probably first evolved not long after the early ancestors of insects started exploring mini-forests of moss and lichen 450 million years ago. Countless species have since independently discovered how useful this way of moving can be.

The building blocks for a really great jump are:
1) long legs that can extend far backwards and so push an animal's body very quickly upwards
2) some material at the top of those long legs which can store lots of pent-up energy that can be released very quickly

And many different types of animals have come up with both of these!

↓ When in danger, **frogs** can power themselves up into the air in a flash. A lot of this power comes from the tendons in their legs, which act like stretched elastic bands. The South African sharp-nosed frog can jump over three metres in a single bound. That's about twice your height!

The **dog flea** (which only likes to live in the warm coat of an unsuspecting pooch) can jump more than *fifty* times its own body length! It does this by storing energy in a pad of super-material in its legs called "resilin", which acts like an incredibly stretchy elastic band that never breaks.

↑ The pretty ordinary-looking bug **Issus coleoptratus** has solved a complicated problem: what happens if one of your legs is stronger than the other? In other jumping insects, if one of their legs pushes off too soon or more powerfully, they are sent spinning around — but not *Issus coleoptratus*. Right at the top of the planthopper's legs are some of the smallest sets of gears in the world. As one leg pushes off the ground, the gears move and push the other leg exactly the right amount. So not only does *Issus coleoptratus* jump long distances — it jumps straighter than most other animals too.

The marsupials of Australasia (mammals like koalas and their relatives that usually have pouches to carry their young) evolved hopping around fifteen million years ago, when the forests of Australia began changing into vast, open grasslands. This meant these creatures couldn't rely on hiding in forests to escape predators and had to spring away on their hind legs instead! **Kangaroo** feet only have four toes, and three of them are very small. When hopping, the large fourth toe pushes off from the ground as the stored energy in the kangaroo's tendons is released, hurtling them forwards at over 60 kilometres per hour.

Living Underground

Climbing or flying are both fantastic ways to avoid predators and to discover more food. But some animals have evolved to head in another direction: downwards.

Burrowing under the ground is a foolproof way to stay out of sight of predators and to avoid cold winters, plus there are tasty worms and grubs to be snaffled. But it isn't an easy way to live. In fact, it tends to require an entire set of super skills. And if you take a look at animals that live underground (or **FOSSORIALLY**), you'll start recognizing these skills everywhere. Burrowing animals need incredibly strong arms for pushing heavy or wet soil or sand out of their way,

Golden mole

↑ In Africa, **golden moles** evolved strong arms and large, clawed, shovel-like hands. But their arms move upwards and downwards rather than from side to side, as in the talpid moles from Asia, Europe and the Americas. Amazingly, they are more closely related to elephants than they are to the talpid moles!

Notoryctes

← Thousands of kilometres away, mammals in Australia, which have been separated from European, African and American mammals for the last 30 million years, also evolved their own mole, on their separate branch on the Tree of Life! *Notoryctes* is a relative of kangaroos and wallabies, and has a backwards-facing pouch so it doesn't fill up with sand when tunnelling.

amazing senses of touch and smell to make up for the fact that eyes are completely useless in the dark, and the ability to breathe when there's hardly any oxygen in those tight, claustrophobic tunnels. Although all these animals survive with the same — or really, really similar — super skills, the way they have evolved is different.

Moles that live in Asia, Europe and the Americas, called talpid moles by zoologists, can push huge amounts of soil up onto the surface as they burrow, thanks to their enormous, spade-like hands. In order to make their hands as big as possible, European moles have a second "thumb" on each hand that extends out of one of their wrist bones.

Talpid mole

Mole cricket

← It's not just mammals that live underground! Chunky, nocturnal **mole crickets** use their large forepaws to dig in a very similar way to moles. What a super skill!

About 150 million years before any of these other animals evolved, the very early mammal *Docofossor* was already living a mole-like lifestyle under the feet of the dinosaurs. Docofossor looked really similar to golden moles, but was only distantly related to pretty much any mammals alive today. ↓

Docofossor

Speedy Swimming

Sleek, speedy ocean predators have evolved many times throughout history, but moving forwards at great speed isn't the easiest thing to do when you are surrounded by water. In fact, it's a real drag.

Sea creatures that swim quickly have very powerful tails and fins to push through the water, but all that water sticks to animals' skin and pulls them backwards. The way water slows down moving animals is called <u>drag</u>. If you want to be fast, it helps to have a few tricks to make your body slip through the water more easily or, to use a fancy word, to be more **HYDRODYNAMIC**.

You need to be very smooth, have a pointed front, a powerful, slim behind, and a stiff tail that moves strongly from side to side or up and down while your front end doesn't move too much. The fancy name for this is thunniform swimming — and it is so effective it's evolved many times in animals that have become superstars of the oceans.

↑ Two-metre-long **bluefin tuna** use a pointed front, a middle packed with muscle and a rigid tail to power through the sea at speeds of nearly 70 kilometres per hour. In fact, thunniform swimming is named after tuna, which are part of a group of fish called the "Thunnini".

→ **Dolphins** evolved a similar hydrodynamic shape to help them catch fast-moving prey. But the reptilian **ichthyosaurs** beat the dolphins to it by about 200 million years! The smooth streamlining, pointed, tooth-filled mouth and stabilizing dorsal fin all helped these two completely different animals to do the same thing: to move fast and catch slippery prey!

Dorsal fins, crescent-shaped tails and smooth, muscular bodies had been striking fear in the deep ocean way before dolphins evolved. In fact, **sharks** evolved this body shape over 100 million years before ichthyosaurs even began to evolve, and they are still around today. Sharks look quite similar to when they first evolved, so this body shape must be a really useful one to have stuck around for such an incredibly long time! There's no doubt they are successful hunters: the slender mako shark can hit speeds of 74 kilometres per hour with a flick of its powerful tail, making it a deadly predator. ↓

← Although not a thunniform swimmer, penguins (like this **gentoo penguin**) share a similar streamlined body shape with these animals. They propel themselves with their wings, but their powerful bodies and sharp, cutting beaks allow them to rip through the water at over 36 kilometres per hour to catch fish.

Dorsal fin

Dolphin

Ichthyosaur

Drinking Nectar

Food is everywhere — if you know where to look and how to reach it.

One particular type of food is found wherever there are flowers, and for millions of years it's been eaten by animals looking for a sugar hit: **NECTAR**.

Nectar is an amazing food that contains sugars and other useful stuff like proteins and minerals that are very important to animals. Most plants produce nectar somewhere inside their flowers, which is a very clever trick. As animals try to reach the nectar, they brush against parts of the flower that produce or receive pollen. When they move to another plant, the animals carry this pollen between flowers, "fertilizing" them so they can grow seeds for the next generation of plants. The animals get a meal, and the plants can grow seeds. It's a fantastic win-win situation called <u>mutualism</u>.

In order to make sure pollen gets rubbed on the animals, many flowers are often awkward shapes. This means that lots of different animals have evolved long, strange-shaped noses or snouts to help them reach the sugary nectar.

Hummingbird

➔ Most animals that feed on nectar are flying insects, like the tens of thousands of species of **bees** and **flies**. Usually, many different kinds of insects can get to a plant's nectar, but some plants allow only a few types of animal to reach their sugary drink. This is so that their pollen is only transferred to another flower of the same species, and not wasted on the wrong kind of plant. An orchid that grows on the island of Madagascar, called *Angraecum sesquipedale*, is one of these. The only insect that can reach the nectar at the end of its thin, long flowers is **Morgan's sphinx moth**, which uses its 30-centimetre-long straw-like mouthpart (its "proboscis") to slurp up the goods!

Morgan's sphinx moth

← The other animals that have excelled at drinking nectar are birds. The most famous are the tiny **hummingbirds** from the Americas, who lap up high-sugar nectar to fuel their high-speed life. They flap their wings more than 80 times per second — that needs a lot of energy!

Honeyeater

Honey possum

↑ Another group of small, long-beaked, nectar-drinking birds also evolved separately in Australasia. **Honeyeaters** don't hover like hummingbirds, but they lap nectar out of flowers in a very similar way. Their tongues trap the liquid, drawing it easily into their throats.

The mammals that feed on nectar are mostly bats, which have very long tongues to reach inside flowers. These include the **grey-headed flying foxes** that feed on the nectar of many Australian plants and, over in Central America, the Mexican long-tailed bat that does the same with a tongue a third as long as its entire body. →

← Apart from bats, there is only one mammal that eats nectar and nothing else. The tiny **honey possum** evolved a long tongue which lets it reach the nectar of flowers in Western Australia.

Grey-headed flying fox

Drinking Blood

Nectar can be found wherever there are flowers, but wherever there are animals, there is another liquid filled with nutritious goodness ... blood.

Drinking blood (or **HEMATOPHAGY** to use the fancy word) is a smart way to feed because, so long as the animal being fed from stays alive, it provides an endless supply of minerals and proteins that gets pumped right into the feeder's stomach. None of that messing about with chewing! But there are a couple of important things animals need to have to drink blood.

First is a very, very sharp mouth to either plunge into an animal's skin, or to snip away a piece of skin to get to the blood underneath. So far, so obvious. But the other thing is much more complicated and it is amazing that it has evolved so many times in different species. It's a cocktail of chemicals (usually found in the animal's saliva) that numbs the area around a wound (so the animal whose blood is being drunk doesn't get annoyed), and that stops the blood from clotting. Blood normally clots at a wound to stop it pouring out, but with these special chemicals vampire species can dine on a flow of lovely, liquid blood until they've had their fill.

The **vampire ground finches** in the Galapagos Islands usually eat a fairly regular diet of fruit and seeds, but, when food is scarce, they become vampires! By pecking at the legs and feathers of birds called boobies, they are able to steal a nutritious meal. Although it must hurt a little, the boobies don't seem to mind vampire ground finches nipping at them that much, a bit like you or I being bothered by mosquitoes. However, they fly away if they're pecked too much. ↓

After discovering the nutritional bonanza that is blood, **calyptra moths** changed their mouthparts at some point in the past from a soft, sucking proboscis for slurping up nectar into a tough, hollow piercing tube for drilling into mammals' skin. Today, male calyptra moths drink blood while females still use their mouthparts to drink nectar and juice from fruit. ↓

Some bloodsuckers don't actually *suck* blood. **Vampire bats** use their sharp front teeth to snip away small parts of skin on large mammals and then lap up the pooling blood with their tongues. ↓

The most famous insects that drink blood are the thousands of species of **mosquitoes.** However, it is only females who ever do this, as they use the contents of the blood to build their eggs. ↓

→ Although **leeches** (a type of worm) are soft and squishy, their mouths are very sharp and perfect for scraping away then digging into the skin of larger animals. To keep their heavy, blood-filled bodies attached to their prey, they hold on with a powerful sucker.

Spines

Animal defences can be fancy and complicated, like spraying venom in the face of an attacker, or using camouflage to blend into the background. But if there's one thing that evolution has shown us time and time again, it's that if something works, there's no need to overcomplicate it!

And that's exactly the deal with **SPINES**. When it comes to something that can keep a predator away, you can't get much simpler than a collection of pointy sharp things aimed at its face. Even large predators like lions need to be *very* careful if they want to nibble on a porcupine.

Wherever spines are found (and they are found on a *lot* of animals), they share two key features: they are sharp and they are tough! The sharpness comes from their shape — skinnier at the tip than at the base. The toughness comes from what they're made out of, which, in mammals, is the super-material keratin (which is also used to build claws, fingernails and horns).

↓ There are seventeen closely related species of **hedgehogs** found across Europe, Africa and Asia. They all protect themselves with a bristling coat of spines that can be pulled over their soft faces and bellies to protect them from owls, foxes and ferrets.

↓ On the island of Madagascar, a separate group of small mammals also discovered the power of prickles. Tenrecs are more closely related to huge underwater manatees than hedgehogs, but they sure look like them! The **lowland streaked tenrec** even uses its spines to communicate, by flicking them against one another to make clear, clicking noises in the undergrowth.

Hedgehog

Lowland streaked tenrec

Old World porcupine

For almost as long as there have been animals on Earth, spines have been protecting creatures under the waves. *Hallucigenia* was a strange, worm-like creature that existed half a billion years ago — before most other kinds of animals even started evolving — and used its spines to protect itself from early predators, like the bizarre-looking *Opabinia*. Today, spines are found all over the oceans, and have evolved in animals like the crown-of-thorns starfish and over 200 species of **pufferfish**.

Pufferfish

Hallucigenia

↓ Rodents have evolved huge spines for protection not once but twice: in the Americas, and again in Africa and India to protect them from lions and leopards. **New World porcupines**, like the North American porcupine, live in trees and sport detachable and very painful spines called quills, with backwards-pointing barbed ends so they stick in any attacker's skin! **Old World porcupines** are active during the night, spend their time on the ground and some, like the crested porcupine, rattle special hollow spines to scare off predators with a warning sound.

New World porcupine

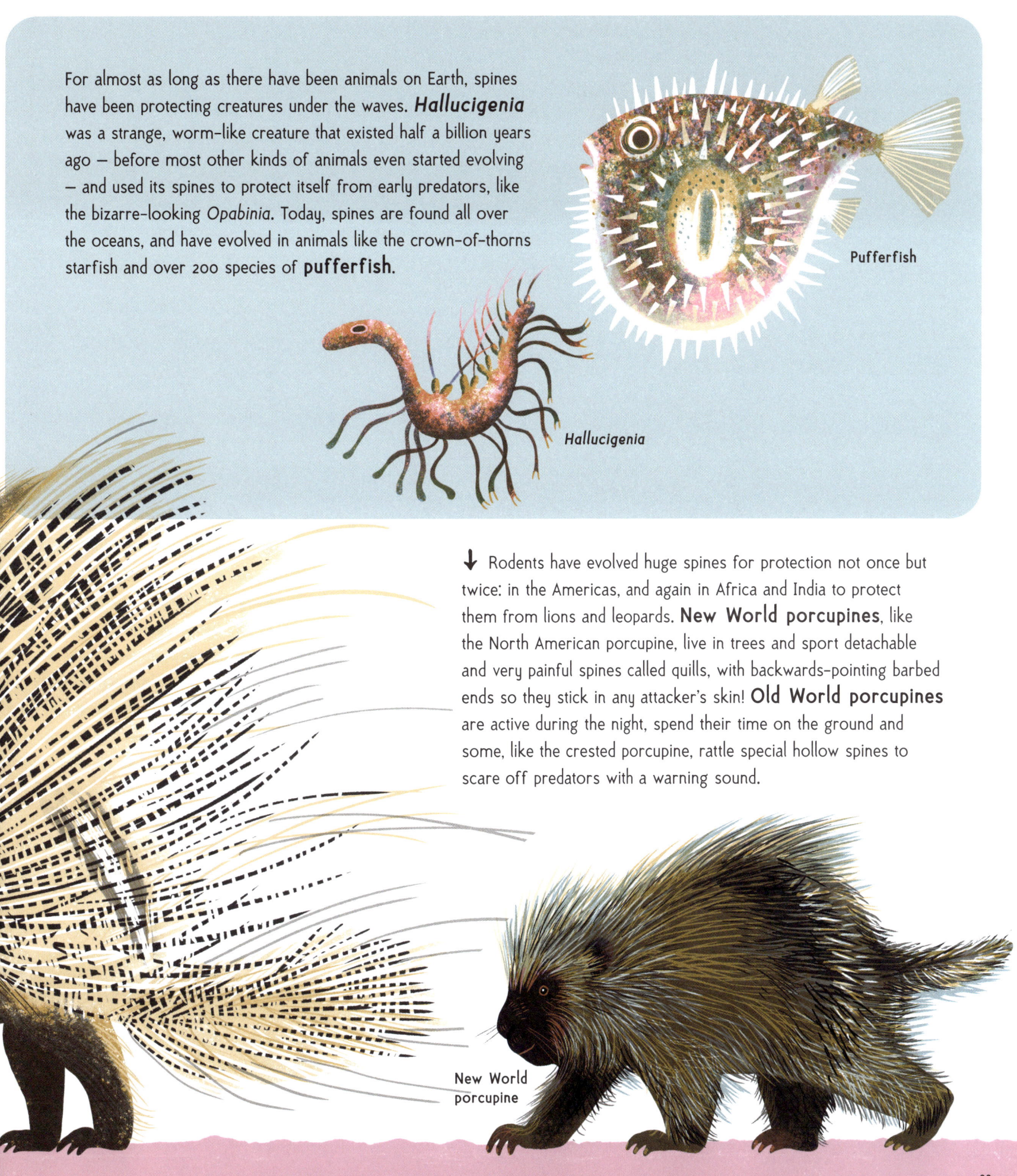

Losing Legs

When we look at animals with backbones today, pretty much everything has legs. Even those animals that don't – like fish – have fins which they use for the same, main purpose: to move!

Although legs seem to be something animals absolutely need, that isn't the case. In fact, **LEGLESSNESS** has evolved lots of times to help animals that live in certain sorts of places move around.

This might seem a little crazy. Why, when nearly everything has legs, would losing them make you a smoother mover? Well, having a very long body makes legs a lot less useful. Lots of animals have evolved long bodies in order to explore small gaps in rocks or under leaves, to search out tasty snacks or find safety. It makes more sense for them to move by pushing their long bodies against the ground, rather than trying to use legs that are very far apart. Losing legs also makes their bodies even sleeker, so they don't get snagged on anything. All of this is why so many animals have ended up legless!

↓ Probably the first legless animals that slither to mind are **snakes**, which evolved from reptiles that had legs. And many early snakes that lived during the time of the dinosaurs, like **Najash** and *Eupodophis*, still had back legs, even though they were so small they wouldn't have been useful for very much! When body parts are used less and less through generations, they become smaller and less important (like the human appendix), and are called "vestigial organs".

Many scientists think that snakes lost their legs because they spent lots of time burrowing, where legs would just get in the way. This makes sense because many other legless animals spend a lot of time in burrows.

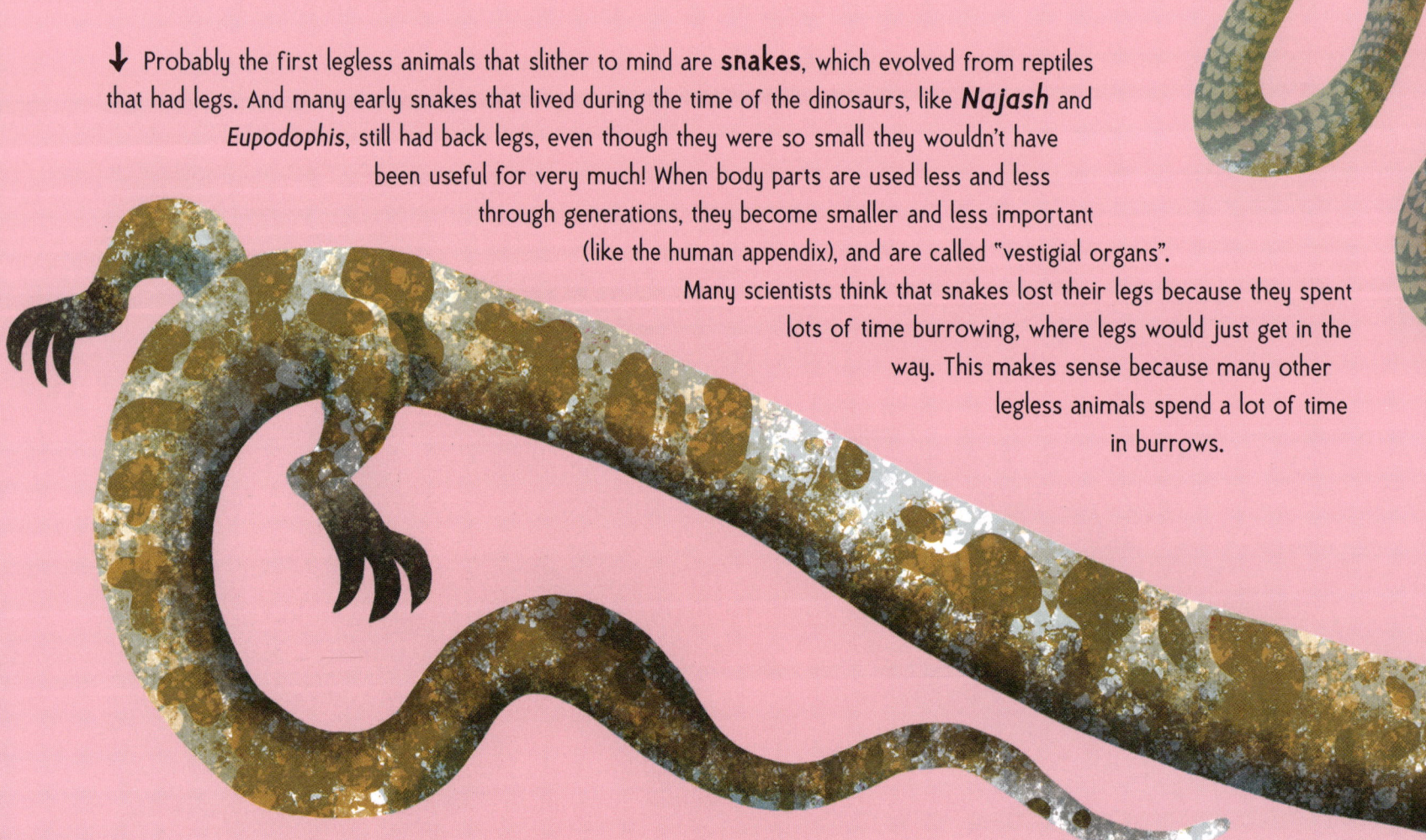

→ Many, many other reptiles also have sleek, legless bodies. Anguids, like **slow-worms** and glass lizards, live in very similar places to snakes and so lost their legs in a similar way to help them get around quickly in the undergrowth. Although they look like them, anguids definitely *aren't* snakes (they have eyelids, which snakes never have). They evolved leglessness completely on their own.

← Some lizards are *almost* limbless, but not quite — perhaps because they haven't had enough time to lose their legs entirely yet. They get along very well with their spike-like legs and may be on their way to true leglessness. Take, for example, **flap-footed lizards** from Australia and New Guinea. They are *very* strange geckos. While many other geckos have feet specialized for climbing, these long lizards spend a lot of time on the ground and have reduced their limbs to only two flap-like back legs, which they barely use. Maybe in a few hundred years, flap-footed lizards won't have any feet, flappy or not!

Foot flap

↓ Way over on another branch of the Tree of Life, amphibians also have their legless group: the worm-like **caecilians**. These smooth sliders live under the ground in damp, tropical places. Their skulls are very strong and pointed, so they can push through the soil and undergrowth to hunt out insects and other invertebrates.

Long Tongues

The trouble with animals evolving ingenious ways to escape danger is that this is *really annoying* for predators. A hunter's gotta eat!

One of the ways that predators have evolved to catch prey is by developing the ability to run after them: gazelles might run quickly, but cheetahs can run even faster... It's an evolutionary game of catch-up, the prey trying to keep one step ahead of the predators.

But for predators whose bodies wouldn't adapt easily to moving very quickly, evolving to run faster just isn't an option. They have to find another way. One solution that has evolved many times in slow-moving predators does involve turning up the speed dial, but just on one part of their bodies: their tongues! These animals have evolved very special muscles called projector muscles, along with elastic body parts that make their tongues **BALLISTIC** (literally meaning they shoot out of their mouths). So even if these predators can't get too close to their prey, they can still pose a supercharged threat...

Projector muscles

Retractor muscles

↓ This *explosive* skill is found in **chameleons**, a group of wacky-eyed, colour-changing hunting lizards found in Africa and parts of Europe and Asia. Their tongues are usually about twice the length of their bodies, and the smallest species, like the **rosette-nosed chameleon**, have the fastest tongues, accelerating to speeds of up to 96 km in only one hundredth of a second!

← Today, there are only three sorts of amphibians: the frogs and toads; the newts and salamanders; and the caecilians. But over 160 million years ago, there was a fourth group — the **albanerpetontids**. These ancient amphibians looked a bit like salamanders and evolved a ballistic tongue all on their own.

→ All **frogs** alive today evolved from frogs that flicked their tongues out relatively slowly to catch prey. Super-fast, ballistic tongue-shooting then evolved many times in frogs, along with the very complicated nerves and brains that are needed to control such fast-moving mouthparts.

→ Many **lungless salamanders** have ballistic tongues to catch prey. Like frogs, these amphibians live in cool, damp places, and evolved ballistic tongues completely on their own. Usually, salamanders whip their tongues out of their mouths (along with the bones at the base of their tongues!), but the lungless salamander's supercharged ballistic tongues are shot out by special spring-like muscles *behind* the tongue. This evolved not once but twice in two different groups of these particularly cute pocket-sized predators.

Prehensile Tails

Walking around on the ground is a pretty safe way to move about: if you trip, you don't have that far to fall. But if you live up in the trees, it's another story.

Lots of animals have taken to the trees to get at tasty fruits and bugs, and it's an excellent way to avoid most of the predators below. But slipping or missing a branch at such great heights could be deadly. So wouldn't it be great for animals up in the trees if they could hold on using <u>five</u> limbs rather than four?

Although the way that bodies grow means it's not really possible for animals with backbones to grow an extra arm or leg very easily, a few animals have all come up with the same solution — they use their tails.

Most animals with backbones have tails, which are mainly used to help them move (like in fish) or for signalling (like in lemurs and lots of species of lizards). However, moveable, wrappable, **PREHENSILE TAILS** have evolved in a host of creatures to help them hold on more tightly in precarious situations.

← Two hundred million years before any other animals evolved prehensile tails, a peculiar bunch of reptiles got there first. **Drepanosaurs** lived during the Triassic period (252–201 million years ago, just when the dinosaurs were beginning to make their mark on Earth), and hung from branches using a claw-tipped tail. Although palaeontologists aren't sure quite how it lived, its sturdy, quick-moving neck and chameleon-like hands seem to show it hunted small, fast-moving prey like insects.

→ The fantastic-looking **binturong** is more closely related to tigers than to monkeys. The last part of its bushy tail is prehensile, which it independently evolved while looking for berries, nuts and insects in the trees of South East Asia. Such a fluffy tail is also an excellent pillow for snuggling into at night.

↑ When we think of animals that are famous for hanging around from their tails, we usually think of monkeys. But only a few types — like spider monkeys and **howler monkeys** — actually have prehensile tails. They are only found in Central and South America. The ends of their tails are bald and sensitive to touch, like the palms of our hands, so they can truly feel and grasp with them!

↑ There are over 200 species of **chameleons** and most of them — including all of the larger ones — rely on their prehensile tails. Chameleons use them not only for a helping "hand" when stretching from branch to branch, but also to hold themselves firm when they're shooting out their powerful projectile tongues at prey.

→ It's not just animals in trees that use grabbing tails. **Seahorses** can't swim like other fish because their bodies are so rigid. Instead, they use their flexible, armoured square tail to hold on to slippery seagrasses and other underwater plants so that they can stay in the same place without being carried away by strong currents.

Working Together

Being alive is one big to-do list. Find food, clean your home, look after the babies... Yeesh! It would help if these jobs could be shared around a bit.

Most animals just look after themselves and their own babies, but in a few remarkable species, tens, hundreds or even *thousands* of individual animals all help each other for the good of an entire colony.

Working together is great! Big things like homes can be built quickly and animals can be protected from predators by depending on strength in numbers. This is called **EUSOCIALITY** and it's a super power because it gives little animals a big advantage. Some types of animals (like mammals) are able to grow to massive sizes because of their clever breathing and blood systems and their strong bones. But insects can't, because they wouldn't be able to get enough oxygen inside their bodies, and their tough external skeletons wouldn't work well at large sizes. But by working *together*, huge numbers of little insects act like larger animals — their different jobs are like the different parts of a large animal, some gathering food, some defending against danger. Each team, or "caste", just does one type of job, and so becomes very, very good at it.

What's really incredible is that this way of living has turned up not only in a bunch of insects, but also in a few other, more surprising, animals!

Loads of species of bees, wasps and ants (which are actually just wasps that have lost their wings) live eusocially and divide jobs between themselves. Some of these ant colonies can be mind-blowingly enormous — **leafcutter ants** can live in colonies of fifteen *million* individual ants — that's *twice* the number of people that live in Hong Kong!

↓ Ants are split into different castes. Queens produce eggs, workers do a lot of building and soldiers protect the colony from intruders. These castes are also found in **termites**. Termites are actually cockroaches, and aren't very closely related to ants, wasps and bees at all.

Queen

King

Soldier

Worker

↑ Translucent **snapping shrimps** are the only known animals that live eusocially in the oceans. Hundreds of these shrimps — all related to each other — live in and around sponges in coral reefs as colonies, protecting their home, foraging for scraps of food and defending their queen with their super-powered pincers.

← Most amazingly, eusociality also evolved in a species of mammal: the bonkers-looking **naked mole rat**. These wrinkly rodents have teeth on the outside of their mouths and are resistant to some sorts of pain and cancer. All of which is interesting enough, but they *also* live in underground colonies of up to 300 rats, with only one female giving birth to the babies. If another female starts behaving like a queen, extremely vicious fights take place to crown the colony leader.

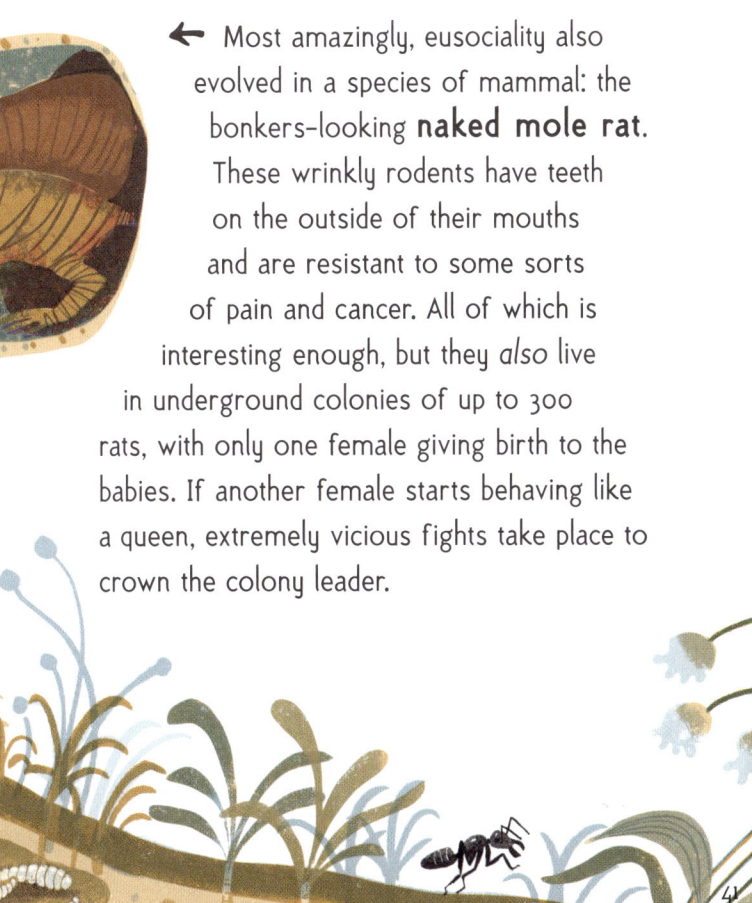

Gliding

Flying is really complicated. To do it you need fantastically powerful arm muscles, precisely shaped wings, finely tuned senses to monitor wind direction and speed, and a whole lot of brain power to keep yourself in the air.

Wouldn't it be wonderful if all the benefits of flying, like escaping predators and travelling distances easily, were available without having to evolve all the body parts needed for powered flight?

Well, animals all over the world have done exactly that by evolving the ability to **GLIDE**.

Gliding is much easier than fully powered flight. All that's really needed is a body part that can be stretched out or flattened to catch the air (called a "patagium"), and the ability to control the direction in which you are gliding. Although the basic idea is the same in all gliding animals, the ways in which their bodies let them do this are all pretty different.

Malayan colugo

Flying squirrel

Wallace's flying frog

→ You might know squirrels as fuzzy-tailed animals that jump from branch to branch above our heads, but there are 50 species of so-called **flying squirrels** in North America, Asia and Europe. These can glide from tree to tree by stretching out the flexible skin that joins their ankles to their wrists.

Volaticotherium

Paradise flying snake

Draco

← Amazingly, very early mammals *totally* unrelated to gliding squirrels and colugos were doing the same thing above the heads of dinosaurs 164 million years ago in the Jurassic period. Mammals like **Volaticotherium** evolved a furry membrane that stretched between their hands, feet and a little bit of their tail, too.

↑ Reptiles have evolved ways to fling themselves across huge distances many times. Today, the **paradise flying snake** in South East Asia flattens out its ribs to slow itself down as it jumps between trees. **Draco**, the "flying" lizards of Sulawesi, glide around Indonesian forests on "wings" of skin draped over extended rib bones.

← Over in South East Asia, the lanky **Malayan colugos** have evolved the same super skill, even though they're more closely related to monkeys than squirrels. They even stretch pieces of skin between their back legs and their tails!

Sharovipteryx

← Like many tropical frogs, **Wallace's flying frog** lives high in the branches of South East Asian forests. Unlike other frogs, when it has to escape from danger quickly, it uses skin stretched between its long fingers and toes to catch the air, helping it parachute away from snakes and birds of prey.

↑ During the Triassic period, the reptile **Sharovipteryx** evolved a slightly different way of gliding. Rather than riding the air with skin held between ribs, *Sharovipteryx* had wings between its back legs and tail. Its triangular patagium worked in a similar way to the wings on delta wing fighter jets built by humans today!

Sabre Teeth

Sometimes evolution has led to some pretty awesome-looking body parts, and it doesn't get much more awesome than the super-long teeth of the sabre-toothed cats.

How exactly sabre-toothed animals used their impressive teeth is still a bit of a mystery. Scientists debate whether the teeth were used like slashing knives, or in a different way. Today, most big cats kill their prey by suffocating them with a strong clamp on their neck, but jaguars use their incredibly strong jaws to bite through animals' skulls! Whether sabre-toothed cats used their teeth like this isn't known, but they did have a whole bunch of other tricks to help them kill prey.

Their mouths could open much, much wider than any modern cat's — over 90 degrees wide in some species! And rather than being slinky and skinny like most cats today, sabre-toothed cats were big, almost bear-like animals with very powerful forearms that could hold down massive prey.

→ If you can picture a sabre-toothed "tiger", then you're probably thinking of *Smilodon*. This famous feline was a member of the Machairodontinae, a group of cats that evolved around sixteen million years ago and used their elongated canine teeth to hunt huge prey such as bison, mammoths and ground sloths. The Machairodonts weren't really tigers, but they were extremely large cats. *Smilodon populator* was probably the biggest cat to have ever lived and could have weighed almost the same as a male polar bear.

↑ ***Thylacosmilus*** might have looked very similar to *Smilodon* but it was more closely related to koalas than kittens. This marsupial that lived around nine to three million years ago evolved huge canine teeth just like the Machairodonts, but with a big difference: those knife-like teeth fitted into sheath-like slots that extended down from its lower jaws.

↑ Over twenty million years before the Machairodonts evolved sabre-teeth, the **nimravids** got there first. These carnivores weren't cats, and were more closely related to modern-day mongoose and hyenas. They had sheaths in their lower jaw for their teeth, just like *Thylacosmilus*.

↑ All of these sabre-toothed predators were late to the party in comparison to a group that were around before the time of the dinosaurs. Two hundred million years before any of these hunters evolved, the **gorgonopsids** had already evolved a pair of terrifying teeth. Gorgonopsids were closely related to the early ancestors of mammals, but lived so long ago that scientists still aren't sure whether they had fur or whether they were cold- or warm-blooded.

The Power of Evolution

The history of life on Earth is a gritty story of incredible challenges, impossible odds and epic struggles to survive. Even today, animals still somehow have to make it through each twenty-four hours without being eaten, or without starving, or without falling out of a tree. In fact, it's amazing so many animals survive!

But, as you've seen, life on Earth is also the story of **INGENIOUS INVENTIONS** and **CRAFTY SOLUTIONS**. The super powers and wonder parts found all over the animal kingdom are the crowning achievements of evolution — the timeless solutions to nature's trickiest problems!

Animals as enormously different as octopuses, guinea pigs and jellyfish have all evolved the same amazing super powers time and time again, over millions of years and on continents separated by thousands of kilometres. It shows not only how brilliantly useful these skills are to animals today, but also how wondrous and efficient evolution is at slowly equipping animals with what they need to eat, hide, hunt and **SURVIVE** on Planet Earth.

Glossary

Evolution
The gradual changes in how a species looks or behaves from one generation to the next through time.

Extinct
When all members of a species have died, that species is said to be extinct. Extinction occurs naturally but can be sped up by the actions of humans on the environment.

Invertebrates
All animals that do not have backbones. Most animals on Earth, from jellyfish to insects, are invertebrates.

Mammals
A group of vertebrates that create milk, have fur or hair on at least some parts of their bodies, and have three small bones inside their ears.

Predator
An animal that hunts other animals (its prey) for food.

Prey
An animal that is hunted by another animal (a predator) for food.

Species
A group of animals that are more like each other than any other animals. Tigers, humans and tuna are all examples of species.

Tree of Life
A way of imagining how all species are related to each other. At the tips of the tree's branches are all the species alive today. The closer the branches, the more closely related the species on the branches.

Vertebrates
Animals that have a backbone, including fish, reptiles, birds, mammals and amphibians.

First published 2022 as *Animal Super Powers: The Most Amazing Ways Animals Have Evolved* by Walker Studio, an imprint of Walker Books Ltd, 87 Vauxhall Walk, London SE11 5HJ

This edition published 2024

2 4 6 8 10 9 7 5 3 1

Text © 2022 Nick Crumpton

Illustrations © 2022 Viola Wang

The right of Nick Crumpton and Viola Wang to be identified as author and illustrator respectively of this work has been asserted in accordance with the Copyright, Designs and Patents Act 1988

This book has been typeset in Polymer

Printed in China

All rights reserved.

No part of this book may be reproduced, transmitted or stored in an information retrieval system in any form or by any means, graphic, electronic or mechanical, including photocopying, taping and recording, without prior written permission from the publisher.

British Library Cataloguing in Publication Data: a catalogue record for this book is available from the British Library

ISBN 978-1-5295-1853-5

www.walkerstudio.com